MY DAD IS A MARINE

Home Is Where My Hero Lives

I0461092

SACRA MAGDALIA DURANDISSE

Copyright

My Dad Is a Marine: Home is where My Hero Lives
Copyright @ 2025 by Sacra M. Durandisse

Barracks Editorial and Design House, LLC.
Orlando, Florida
iambevtheeditor@gmail.com.

Bookcover: AI-assisted, redesigned by Freelance Designer: Ashik: *ashikarosh24@gmail.com.*

Illustrations were produced using AI-assisted digital tools and are the exclusive property of the author.

AI-enchanced digital illustrations, developed under the direction of **Sacra M. Durandisse.**

All Rights Reserved. This book or any portion thereof may not be reproduced or used in any form or manner without the author's express written permission, except for the use of brief quotations in a book review, except as provided by USA copyright law.

Although the author has made every effort to ensure that the information in this book was correct at press time, the author and publisher do not assume and, with this, disclaim any liability to any party for any loss, damage, or disruption caused by errors or omissions, whether such errors or omissions result from negligence, accidents, or any other cause.

For information on booking interviews, book signings, all inquiries, or other events, please get in touch with the author via email at:

sdurandisse@yahoo.com.
ISBN: 979-8-218-89551-8

Printed in the United States of America

My Dad Is A Marine
Home is where My Hero Lives

GIFTED TO:

_____

FROM:

_____

DATE:

_____

ACKNOWLEDGEMENT

This story is inspired by a remarkable single father and Marine,

Officer **Juan Arce,** whom I had the honor of meeting while working in Japan. Over his 23 years of service, Officer Arce has shown strength, dedication, and compassion, qualities that shine just as brightly in his role as a dedicated father.

TRIBUTE

A touching tribute to every child who proudly calls a Marine "Dad."

INTRODUCTION

This is a Heartwarming Children's book about family, courage, and love of country, the USA. Inspired by a remarkable single father and Marine, Officer Juan Arce, whom I had the honor of meeting while working in Japan. Over his 23 years of service, Officer Arce has shown strength, dedication, and compassion, qualities that shine just as brightly in his role as a single father of three beautiful girls.

Through the eyes of a bright and curious little girl, "My Dad Is a Marine" beautifully captures the love, pride, and sacrifice that come with being part of a military family. I watched the bond he shared with his daughters, which inspired me to tell their story through the eyes of his six-year-old daughter. This story also celebrates family, resilience, bravery, and the everyday heroism found in love.

Officer Juan Arce is not just her hero at home; he is a hero to the nation. From early morning salutes to bedtime stories filled with bravery, she learns that being a Marine means serving with courage, strength, and heart.

This inspiring story gently introduces children to the values of discipline, service, and honor, while celebrating the unbreakable bond between father and child.

With gratitude and respect,

Sacra Magdalia Durandisse

My Dad Is A Marine
Home is where My Hero Lives

My name is Marley, and I am six years old. My dad is a Marine. He is big, strong, and my hero!

My dad takes me to the park, gymnastics, and cheerleading practice.

He always cheers the loudest and makes me feel so proud!

My dad cooks dinner for me and my sisters; he's a great cook!

We love it when he makes pizza, and he always lets me help.

Sometimes, I sneak our dog Bentley a little pepperoni snack when Dad isn't looking.

My dad sometimes takes me and my sisters out for dinner.

We love eating *Mexican* food together as a family!

Dessert time is my favorite, especially when I get ice cream.

Vanilla is the Best!

My dad helps me ride my bike.
When I fall, he picks me up and gives me
a big hug. He kisses my boo-boo and says,
"You have to balance yourself and stay
straight on your bike. You can do it! You're
strong, just like me."

My dad drives me to school every morning
and picks me up after.

He waits in the long car line, just so he
can see me first when the bell rings.

My dad takes me and my sisters on vacation.

We love traveling on airplanes with him!
We have been to *Hawaii, Thailand, Malaysia, Singapore, Korea, Japan,* and even *Nicaragua!*

We love going to the beaches in Nicaragua,
they are so beautiful!
I love feeling the cool water on my feet
and playing in the soft, warm sand.

We hiked in *Hawaii,* saw beautiful temples in *Thailand,* and visited historic sites in *Japan.* My older sisters even went scuba diving in *Japan;* they were so brave!

In every country we visit, we explore the local markets and buy little trinkets to remind us of our adventures together.

We have also visited many states in *America, like California, Maryland, Pennsylvania, Virginia, Washington, D.C., Connecticut,* and *Florida.* I love the beaches in California and *Florida!*

Those two states have Disneyland in *California* and Disney World in *Florida,* and my favorite ride is the Merry-Go-Round!

We love going fishing in Florida and spending time with our family.

In Washington, D.C., we saw the White House, the Lincoln Memorial, the Washington Monument, and lots of museums.

We learned so much! There were tall buildings, museums, and statues everywhere.

When we visit different countries and states, my dad always makes sure we learn about the places we go.

Sometimes, my dad takes me to work
with him on the base.

I get to sit in his office and draw pictures
while he works.

It's so much fun!

My dad always salutes the *United States* flag. He says he loves his country, and that's why he became a Marine.

My dad is brave. He has fought in wars to keep me, my sisters, our family, *America,* and the world safe.

My sisters told me that when they were my age, my dad had to go away for a very long time.

They said they missed him every single day. Sometimes they cried when he had to leave, wishing he could stay just a little longer.

But when he finally came home safe, they ran into his arms and held him tight; they didn't want to ever let go.

My dad tells us the coolest stories ever!

He has flown in helicopters, ridden in tanks, and even flown in giant military airplanes!

One time, he even took us on a real

helicopter ride!

It felt like we were flying right up into the

clouds. I felt like a superhero!

My dad is the best!

Even after long days working on the base,

he always makes time for me.

He takes care of me, helps me with my

homework, and even washes and combs my

hair.

He lets me have playdates with my

cousins and friends.

We have so much fun together!

My dad loves to have fun and entertain his friends.

His Marine buddies come over to barbecue and tell funny stories.

They work hard to keep everyone in America safe and free.

My dad always smiles and laughs when he's with his friends; they're like one big Marine family.

We love that our dad is a Marine, but
sometimes it's not easy.

We move a lot, sometimes every two to
four years.

Each time, we have to pack up our things,
start a new school, and make new friends.

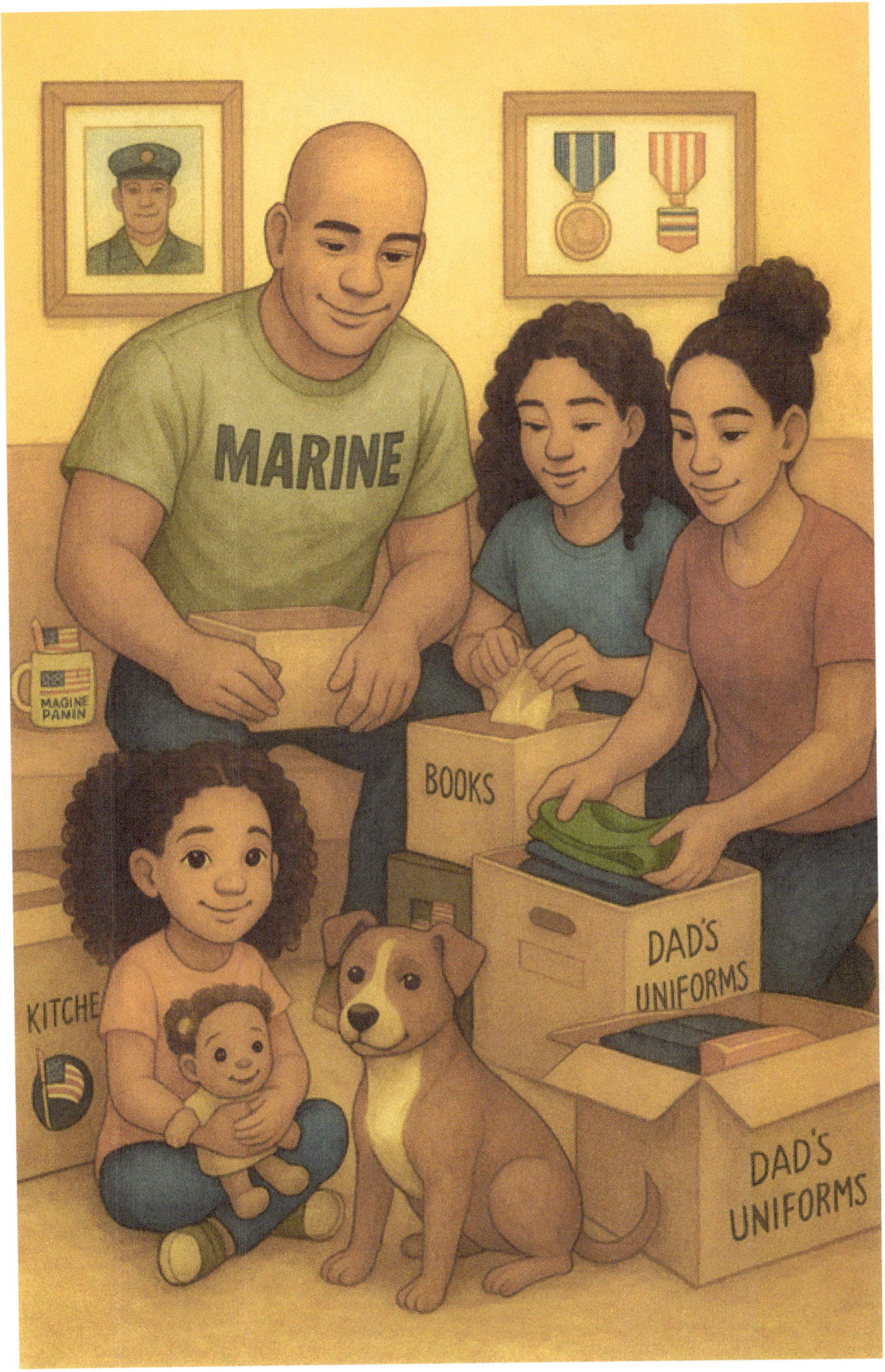
MARINE
MAGINE PAMIN
BOOKS
DAD'S UNIFORMS
KITCHE
DAD'S UNIFORMS

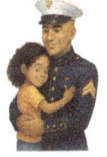

It is always hard to say goodbye, but Dad reminds us that adventure and new memories are waiting wherever we go.

My dad has a way of making everything easier to understand and accept.

He always says, "We're family, and as long as we have each other, we'll be just fine, no matter where we are."

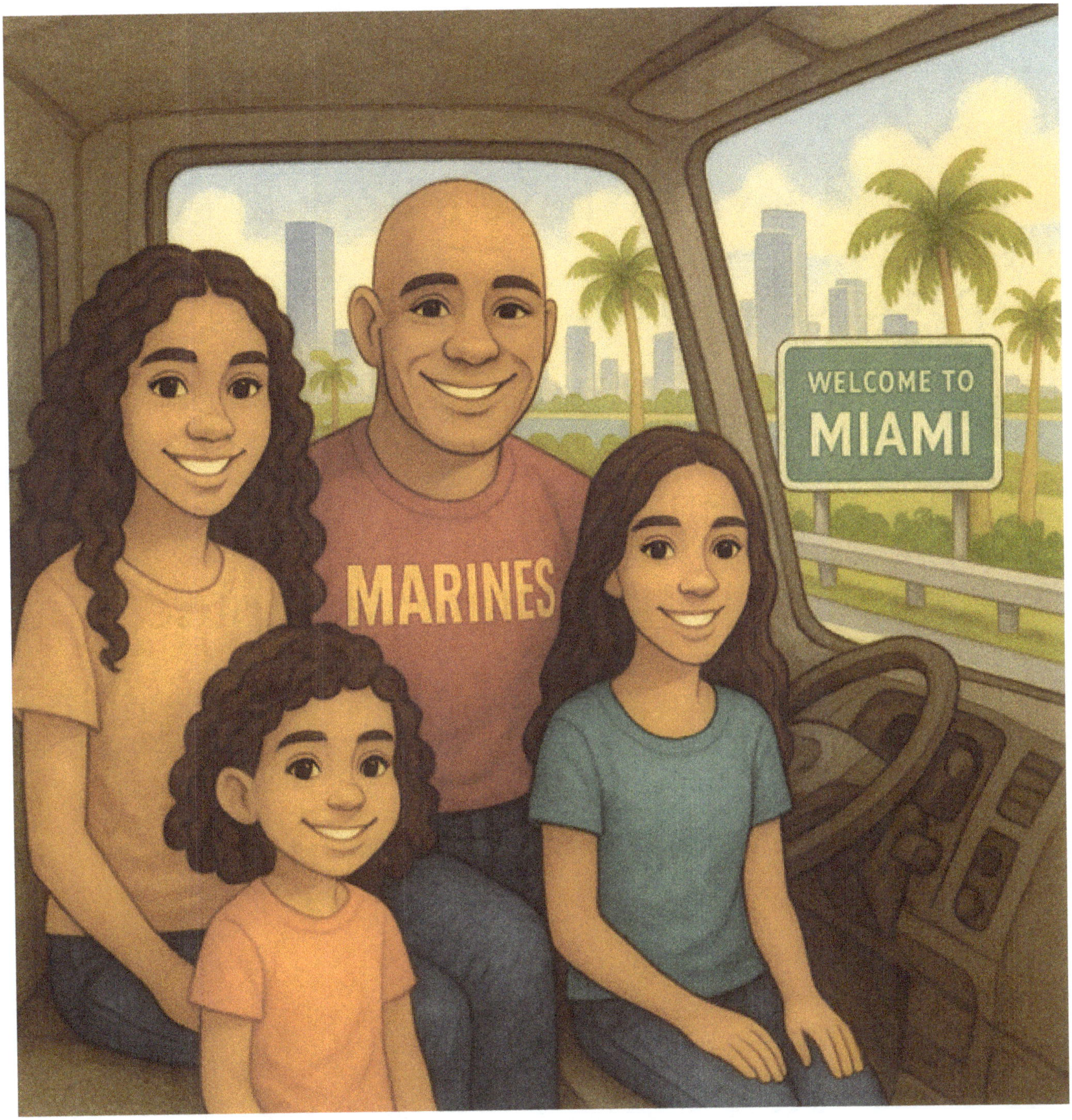
WELCOME TO
MIAMI
MARINES

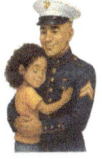

My dad says we have to look at every experience as an opportunity and a new adventure.

He says many people never get the chance to live in the different countries and states we've had the opportunity to call home.

I love my dad very much. My dad loves
me and my sisters very much.
He is a Marine, and he protects me and
our family. I am so proud of him.

The End.

Letter to My Hero

To every Marine, soldier, and service member, thank you for your courage, your sacrifice, and your love.

This story is a celebration of the families who serve alongside you, the children who wait proudly, and the love that never fades.

With gratitude and respect,

Sacra Magdalia Durandisse

Letter to my Hero:

_____

_____

_____

_____

_____

_____

_____

_____

_____

_____

About the Author

Sacra M. Durandisse is an accomplished mental health professional, author, and proud supporter of military families. With nearly three decades of experience in behavioral health and leadership, she has dedicated her life to helping others heal, grow, and find strength through love and resilience.

Inspired by real families she's met around the world, including the brave and devoted Marine who inspired My Dad Is A Marine, Sacra brings to life stories that celebrate courage, family, and hope. Her writing is filled with compassion and a deep respect for those who serve and the families who stand beside them.

When she's not writing, Sacra enjoys traveling, mentoring young professionals, and spending time with her loved ones. She hopes this story inspires children everywhere to take pride in their families, honor their heroes, and always believe in the power of love and togetherness.

www.ingramcontent.com/pod-product-compliance
Lightning Source LLC
Chambersburg PA
CBHW082113120626
46553CB00011B/3656